KV-681-790

The Open University

Mathematics: A Second Level Course

Linear Mathematics   Unit 32

THE HEAT CONDUCTION EQUATION

Prepared by the Linear Mathematics Course Team

CHRIST'S COLLEGE
LIBRARY

The Open University Press

Quarto
512.897

The Open University Press Walton Hall Bletchley Bucks

First published 1972

Copyright © 1972 The Open University

All rights reserved. No part of this work may
be reproduced in any form, by mimeograph
or any other means, without permission in
writing from the publishers.

Designed by the Media Development Group of the Open University.

Printed in Great Britain by
Martin Cadbury Printing Group

SBN 335 01122 5

CHRIST'S COLLEGE
LIBRARY

Accession No.	49146
Class No.	Q512.897
Catal.	− 4/1975

This text forms part of the correspondence element of an Open University
Second Level Course. The complete list of units in the course is given at
the end of this text.

For general availability of supporting material referred to in this text,
please write to the Director of Marketing, The Open University, Walton
Hall, Bletchley, Buckinghamshire.

Further information on Open University courses may be obtained from
the Admissions Office, The Open University, P.O. Box 48, Bletchley,
Buckinghamshire.

Contents

Set Books

D. L. Kreider, R. G. Kuller, D. R. Ostberg and F. W. Perkins, *An Introduction to Linear Analysis* (Addison-Wesley, 1966).

E. D. Nering, *Linear Algebra and Matrix Theory* (John Wiley, 1970).

It is essential to have these books; the course is based on them and will not make sense without them.

Conventions

Before working through this correspondence text make sure you have read *A Guide to the Linear Mathematics Course*. Of the typographical conventions given in the Guide the following are the most important.

The set books are referred to as:

> **K** for *An Introduction to Linear Analysis*
> **N** for *Linear Algebra and Matrix Theory*

All starred items in the summaries are examinable.

References to the Open University Mathematics Foundation Course Units (The Open University Press, 1971) take the form *Unit M100 3, Operations and Morphisms*.

32.0 INTRODUCTION

In *Unit 23, The Wave Equation*, we have seen how to use the one-dimensional wave equation in modelling the vibrations of mechanical systems such as the air in a musical instrument. The wave equation is only one of a variety of partial differential equations which are of great value in mathematical modelling. The intention of this unit is to introduce a few of these other equations. As a case study for the use of these equations in modelling, we shall take the conduction of heat in a solid. We shall see how different physical situations lead to different equations (although these are all special cases of one very general equation), and how equations that are superficially very similar may have very different properties.

We shall begin by showing how physical problems involving heat conduction can be modelled by partial differential equations with suitable boundary conditions. We shall then show how some of these boundary-value problems can be solved using extensions of the technique of separation of variables treated in *Unit 23*. Finally we shall compare the various equations with each other and with the one-dimensional wave equation, and indicate some of the ways in which the theory generalizes to more complicated cases.

We shall concentrate on three equations:
the one-dimensional heat conduction equation

$$\frac{\partial^2 u}{\partial x^2} = a^2 \frac{\partial u}{\partial t}$$

the two-dimensional Laplace equation

$$\frac{\partial^2 u}{\partial x^2} + \frac{\partial^2 u}{\partial y^2} = 0$$

and the version of this equation used when the rectangular coordinates x, y in the plane are replaced by polar coordinates r, θ (see sub-section 27.1.2 of *Unit M100 27, Complex Numbers I*)

$$\frac{\partial^2 u}{\partial r^2} + \frac{1}{r}\frac{\partial u}{\partial r} + \frac{1}{r^2}\frac{\partial^2 u}{\partial \theta^2} = 0.$$

CHRIST'S COLLEGE LIBRARY

5

32.1 THE FORMULATION OF PARTIAL DIFFERENTIAL EQUATIONS

32.1.1 The One-dimensional Heat Conduction Equation

Two of the principal steps in setting up a mathematical model are:

(i)   specify the unknown quantities;

(ii) formulate the scientific laws connecting them.

In problems concerning the conduction of heat through a solid the unknown quantity we wish to calculate is the temperature. It will depend in general on position and time and will therefore be a function u of three space variables x, y, z and one time variable t

$$u: (x, y, z, t) \longmapsto u(x, y, z, t).$$

In this unit, however, we shall simplify the discussion by confining it to cases where the temperature depends only on two of these variables, either x and t or x and y, so that

$$u: (x, t) \longmapsto u(x, t)$$

or

$$u: (x, y) \longmapsto u(x, y).$$

The scientific laws relating to heat conduction are considered in the next reading passage, where they are used to formulate the differential equation of heat conduction for a one-dimensional problem.

READ from line 9 on page **K**512 *to line 5 on page* **K**513; then

READ from line 8 on page **K**514 *to the end of the section.*

Notes

(i)   *line 14, page* **K**512 "constant on each cross section". That is, the temperature in the rod is independent of y and z and can therefore be described by a function

$$u: (x, t) \longmapsto u(x, t)$$

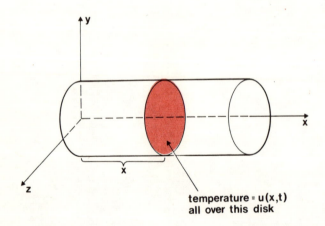

temperature = u(x,t)
all over this disk

(ii) *line 18, page* **K**512   The "rate of change" is a *distance* rate of change.

(iii) *Equation (13-10), page* **K**512   This is the law of heat conduction, sometimes called *Fourier's Law*, and is the first of three empirical scientific laws used in formulating the heat conduction equation. Here ΔH does not mean "Δ times H" but is a single symbol. Since this symbol is used in a different sense in the next equation in **K**, it may be clearer to re-write Equation (13-10) as

$$J = -k \frac{\partial u}{\partial x}$$

where J means the amount of heat crossing the disk (shaded in the above diagram) per unit time. A positive sign for J indicates that the heat is flowing to the right.

The number k is proportional to the area of this disk, that is to the area of cross-section of the bar. The constant of proportionality (the value of k for a cross-section of unit area) is called the *thermal conductivity* of the material of the bar.

(iv)   *Equation (13-11), page* K512   This is the second of the three empirical laws: the amount of heat supplied to a body equals the specific heat times the mass times the rise in temperature. The symbol ΔH now has a different meaning from that given to it in Equation (13-10). It stands for the rate at which heat is accumulating in some portion of the rod whose mass is m and whose length is short enough to permit us to treat the temperature as uniform (i.e. independent of x) along it. This approximation does not lead to any error in the final result (as we shall see in note (vii)).

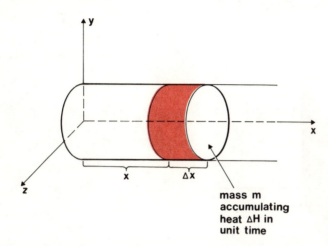

mass m
accumulating
heat ΔH in
unit time

(v)   *line* -7, *page* K512   The symbol Δx is a single symbol, like ΔH.

(vi)   *line* -1, *page* K512   This equation comes from the third of the three empirical laws, which tells us that (in the physical situation considered here) heat is not created or destroyed; so, the amount accumulating in any region equals the amount coming in minus the amount going out during the same time interval:

accumulation = input − output

Laws of this form are called *conservation laws*. For our problem, the conservation law for heat applied to the portion of the bar between x and $x + \Delta x$ shown in the diagram above, gives

$$\Delta H = J(x, t) - J(x + \Delta x, t)$$

since ΔH is the rate at which heat is accumulating, $J(x, t)$ is the rate at which it is entering the left-hand end of the portion (at time t), and $J(x + \Delta x, t)$ is the rate at which it is leaving the right-hand end.

Combining the above equation with the one in note (iii), we obtain

$$\Delta H = k \frac{\partial u}{\partial x}(x + \Delta x, t) - k \frac{\partial u}{\partial x}(x, t)$$

which is the equation at the bottom of page K512, expressed in the notation of *Unit 23*.

(vii)   *lines 2, 3 and 4, page* K513   In the notation used just above, this equation is

$$\frac{1}{\Delta x}\left(\frac{\partial u}{\partial x}(x + \Delta x, t) - \frac{\partial u}{\partial x}(x, t)\right) = \frac{c\rho}{k}\frac{\partial u}{\partial t}(x, t)$$

The reason for taking the limit of small Δx is that it eliminates the error due to the approximation mentioned in note (iv), where we assumed that the temperature at time t was the same throughout the portion. (A similar limit process is used in *Unit 23* to derive the wave equation.) In this limit the left-hand side of the above equation approaches

$$\frac{\partial}{\partial x}\left(\frac{\partial u}{\partial x}(x, t)\right),$$

by the definition of a partial derivative, and so we obtain line 4 on page K513.

(viii)   *line 11, page* K514   The symbol A stands for the x-coordinate of an end of the rod.

7

(ix)   *line 15, page* K514 "insulated at A".   The vanishing of $\frac{\partial u}{\partial x}(A, t)$ in this case follows from Equation (13-10).

(x)   *last line of the section, page* K514   If one end of the rod is exposed to the air, it is a fair approximation that the rate of heat flow out of it is proportional to the difference in temperature between the rod and the air. Thus, to use boundary condition 3 we should take the zero of temperature measurement as the temperature of the surrounding air.

In the derivation given, the rod is assumed to be slender and surrounded by insulating material. Would the same equation apply if the rod were not slender, e.g. if it were replaced by a slab whose two faces were each at uniform temperatures (i.e. temperatures independent of y and z)?

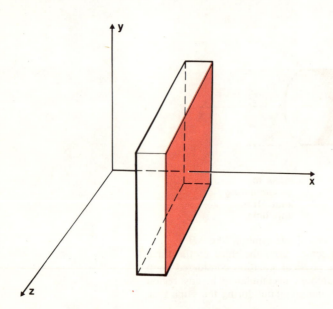

In fact, precisely the same equation applies. It does not even matter very much if the edges of the slab are not insulated, since they are too far away from most of the slab to have a significant effect. The need to consider a thin rod in the previous derivation was to ensure that the temperature distribution is constant over any cross-section. For the slab, such a distribution is ensured by maintaining each large face at uniform temperature.

Exercises

1.   In the boundary condition 3 on page K514, which sign should the number h have at (i) the right-hand end of the rod, (ii) the left-hand end, (assuming x increases from left to right) to model the physical situation described in note (x) above?

2.   Exercise 2, page K514.

Solutions

1.   If the temperature of the rod exceeds that of the surrounding air, i.e. $u(x, t) > 0$, then heat will flow out of it. At the right-hand end (case (a)) such a flow corresponds to $J > 0$ (where J is defined in note (iii)) and hence to $\frac{\partial u}{\partial x}(A, t) < 0$. Since u and $\frac{\partial u}{\partial x}$ have opposite signs at (A, t) we need $h < 0$ in this case. At the left-hand end (case (b)) heat flowing out of the rod is travelling to the left, so that $J < 0$, $\frac{\partial u}{\partial x}(A, t) > 0$, and we need $h > 0$.

2. This time the principle of conservation of heat must be modified to

$$\text{accumulation} = \text{input} - \text{output} + \text{production}$$

$$\Delta H = kJ(x) - kJ(x + \Delta x) + g(x, t)\,\Delta x$$

Using the first two empirical laws, we obtain

$$c\rho\,\Delta x\,\frac{\partial u}{\partial t}(x, t) = k\left(\frac{\partial u}{\partial x}(x + \Delta x, t) - \frac{\partial u}{\partial x}(x, t)\right)$$
$$+ g(x, t)\,\Delta x.$$

Dividing by Δx and taking the limit of small Δx then gives

$$c\rho\,\frac{\partial u}{\partial t}(x, t) = k\,\frac{\partial^2 u}{\partial x^2}(x, t) + g(x, t).$$

A further division by k gives the equation in the question, with $\dfrac{1}{k}$ for b, and $\dfrac{c\rho}{k}$ for a^2.

32.1.2 Laplace's Equation

To show how the discussion given in the preceding sub-section can be generalized, we now derive an equation describing heat flow in two dimensions. A physical situation leading to such a problem is a metal plate of any shape, say a circle or rectangle, sandwiched between two layers of insulating material, so that heat does not flow perpendicular to the faces, and the temperature is therefore independent of the co-ordinate in this direction, but depends on the other two. In the situation depicted in the diagram the temperature depends on x and y, but not on z.

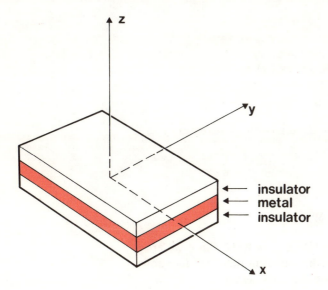

To keep the number of variables down to two, we shall assume in our derivation that the temperature does not depend on time. (In other words the slab has been heated up in some way and has then been allowed to settle down to some "steady" temperature distribution.) The temperature function then has the form

$$u: (x, y) \longmapsto u(x, y)$$

with domain corresponding to the region of the xy-plane occupied by the plate. Such a temperature function, independent of the time, is said to describe a *steady-state* situation.

To formulate the equation we use the same empirical laws as before, but instead of applying them to a portion of a rod with length Δx we now take a rectangular portion of the plate.

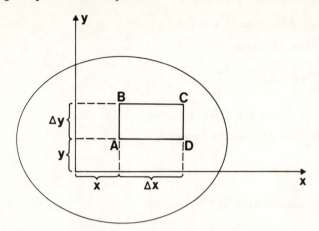

(i) Fourier's law relates the heat flows across the sides of the portion to the derivatives of $u(x, y)$. Consider the side AB. An approximation to the temperature derivative in the x-direction at points along AB is $\dfrac{\partial u}{\partial x}$ evaluated at the mid-point of AB. If we write J_{AB} for the heat crossing side AB to the right, we then have

$$J_{AB} \simeq -k\frac{\partial u}{\partial x}(x, y + \tfrac{1}{2}\,\Delta y)$$

where k is a constant of proportionality analogous to the one used in the preceding sub-section (Equation (13-10), page K512). The heat crossing AB for a given $\dfrac{\partial u}{\partial x}$ is proportional to the length of AB; so k will also be proportional to this length, that is to Δy. Calling the constant of proportionality k_1, we thus have

$$J_{AB} \simeq -k_1\,\Delta y\,\frac{\partial u}{\partial x}(x, y + \tfrac{1}{2}\Delta y)$$

In a similar way the heat flow to the right across CD is

$$J_{CD} \simeq -k_1\,\Delta y\,\frac{\partial u}{\partial x}(x + \Delta x, y + \tfrac{1}{2}\Delta y)$$

For the flows across AD and BC we take the direction of increasing y as positive, obtaining

$$J_{AD} \simeq -k_1\,\Delta x\,\frac{\partial u}{\partial y}(x + \tfrac{1}{2}\Delta x, y)$$

$$J_{BC} \simeq -k_1\,\Delta x\,\frac{\partial u}{\partial y}(x + \tfrac{1}{2}\Delta x, y + \Delta y)$$

(ii) The specific heat law is Equation (13-11) on page K512. Since we are considering a steady state, this reduces to the statement that heat does not accumulate in the rectangular portion we are considering.

(iii) The conservation law. Since there is no accumulation, this reduces to

$$0 = \text{input} - \text{output}$$
$$= (J_{AB} + J_{AD}) - (J_{BC} + J_{CD})$$
$$= (J_{AB} - J_{CD}) + (J_{AD} - J_{BC})$$
$$\simeq k_1\,\Delta y\left(-\frac{\partial u}{\partial x}(x, y + \tfrac{1}{2}\Delta y) + \frac{\partial u}{\partial x}(x + \Delta x, y + \tfrac{1}{2}\Delta y)\right)$$
$$+ k_1\,\Delta x\left(-\frac{\partial u}{\partial y}(x + \tfrac{1}{2}\Delta x, y) + \frac{\partial u}{\partial y}(x + \tfrac{1}{2}\Delta x, y + \Delta y)\right)$$

We now take the limits of small Δx and Δy, first dividing both sides of the equation of $\Delta x \times \Delta y$ so as to get a non-trivial result.

$$0 = \lim_{\substack{\Delta x \to 0 \\ \Delta y \to 0}} k_1 \left(\frac{\frac{\partial u}{\partial x}(x + \Delta x, y + \frac{1}{2}\Delta y) - \frac{\partial u}{\partial x}(x, y + \frac{1}{2}\Delta y)}{\Delta x} \right)$$

$$+ \lim_{\substack{\Delta x \to 0 \\ \Delta y \to 0}} k_1 \left(\frac{\frac{\partial u}{\partial y}(x + \frac{1}{2}\Delta x, y + \Delta y) - \frac{\partial u}{\partial y}(x + \frac{1}{2}\Delta x, y)}{\Delta y} \right)$$

For the first term, taking the limit for Δy first (x and y are independent of each other) and then the limit for Δx gives, by the definition of a partial derivative,

$$k_1 \frac{\partial}{\partial x}\left(\frac{\partial u}{\partial x}\right)(x, y).$$

A similar argument for the second term yields

$$k_1 \frac{\partial}{\partial y}\left(\frac{\partial u}{\partial y}\right)(x, y).$$

Thus

$$0 = k_1 \left(\frac{\partial}{\partial x}\left(\frac{\partial u}{\partial x}\right) + \frac{\partial}{\partial y}\left(\frac{\partial u}{\partial y}\right) \right)$$

which is equivalent to

$$\frac{\partial^2 u}{\partial x^2} + \frac{\partial^2 u}{\partial y^2} = 0. \tag{1}$$

This is called *Laplace's equation in two dimensions*.

When we come to solve Laplace's equation (1) by separation of variables, we shall find that it can be done only if the plate is rectangular. This is because the method only works if the boundary conditions separate as well as the differential equation. For this to happen values of x on the boundary must be independent of y, and vice versa. The only way of extending the method of separating variables to other shapes is to use other coordinate systems (where possible). The *polar coordinate system*, which we met in sub-section 27.1.2 of *Unit M100 27, Complex Numbers I*, is useful, for example, in heat conduction problems in pipes of circular cross-section. The polar coordinates (r, θ) of a point in the plane are related to its rectangular coordinates (x, y) by

$$x = r \cos \theta$$

$$y = r \sin \theta$$

One way of obtaining the partial differential equation of heat conduction in polar coordinates is to work directly from the above relations. You will find the calculation on pages **K546** to 547, but the manipulations are rather heavy and of a type that may be unfamiliar to you. An alternative derivation is to work from first principles, as in the preceding sub-section, using

11

the new coordinate system. This derivation is given in the Appendix and yields

$$\frac{1}{r}\frac{\partial}{\partial r}\left(r\frac{\partial u}{\partial r}\right) + \frac{1}{r^2}\frac{\partial^2 u}{\partial\theta^2} = 0$$

or

$$\frac{\partial^2 u}{\partial r^2} + \frac{1}{r}\frac{\partial u}{\partial r} + \frac{1}{r^2}\frac{\partial^2 u}{\partial\theta^2} = 0 \qquad (r \neq 0) \tag{2}$$

which is called *Laplace's equation in polar co-ordinates.**

Exercises

1. Generalize the discussion given above for deriving Equation (1) to the case where the temperature may depend on time, t, as well as x and y.

2. Write down the boundary conditions on $u(x, y)$ that model the following physical boundary conditions at the edges of a metal plate sandwiched between two insulators in the z-direction:

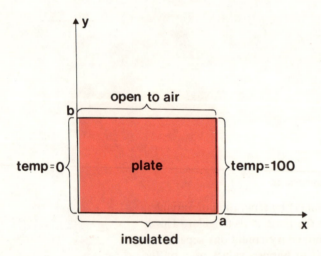

Take the temperature of the air to be 0. If you introduce any constant of proportionality, state its sign.

3. Formulate a differential equation satisfied by the pressure of oil in a porous rock under the following assumptions.

   (a) The problem is two-dimensional.

   (b) The quantity of oil crossing unit length per unit time is proportional to the partial derivative of the pressure function with respect to a coordinate measuring distance at right angles to the unit length.

   (c) Oil is neither created nor destroyed, i.e.
   input − output = accumulation = 0.

Solutions

1. The argument given in the text applies with the following modifications.

   (a) The temperature function is now

   $$u: (x, y, t) \longmapsto u(x, y, t).$$

   (b) Change $u(x, y + \frac{1}{2}\Delta y)$ to $u(x, y + \frac{1}{2}\Delta y, t)$, etc. in the equations for J_{AB}, etc.

* If you found the derivation of Laplace's Equation in Cartesian coordinates difficult, you may find the Appendix very helpful since the same method is employed.

(c)   The specific heat law gives

$$\Delta H = c\rho_1 \, \Delta x \, \Delta y \, \frac{\partial u}{\partial t}$$

where ΔH is the heat accumulating per unit area per unit time and ρ_1 is the mass of the plate per unit area.

(d)   In the conservation law, replace the zero in the left-hand side by $c\rho_1 \, \Delta x \, \Delta y \, \dfrac{\partial u}{\partial t}$. The final result is then

$$\frac{\partial^2 u}{\partial x^2} + \frac{\partial^2 u}{\partial y^2} = a^2 \frac{\partial u}{\partial t}$$

where

$$a^2 = \frac{c\rho_1}{k_1}.$$

This is the *two-dimensional heat conduction equation.* Notice that for steady state problems it reduces to Laplace's Equation, since in those cases $\dfrac{\partial u}{\partial t} = 0$, and that if u is independent of y, i.e. $\dfrac{\partial^2 u}{\partial y^2} = 0$, it reduces to the one-dimensional heat-conduction equation.

2. For the edge on the y-axis: $u(0, y) = 0$
    For the edge on $x = a$: $u(a, y) = 100$ $\qquad (y \in [0, b])$

    For the edge on x-axis: $\dfrac{\partial u}{\partial y}(x, 0) = 0$

    $\qquad\qquad\qquad\qquad\qquad\qquad\qquad (x \in [0, a])$

    For the edge on $y = b$: $\dfrac{\partial u}{\partial y}(x, b) = hu(x, b)$

    where h is a *negative* constant. In the third line we use $\dfrac{\partial u}{\partial y}$, not $\dfrac{\partial u}{\partial x}$ as on page **K514**, because the heat flow that the insulation fixes at zero is in the y-direction. The fourth line corresponds to condition 3 on page **K514**, and the sign of h is as discussed in Exercise 1 of sub-section 32.1.1.

3. The figure shows a rectangular portion of the rock with sides of length Δx, Δy, where x and y are Cartesian coordinates.

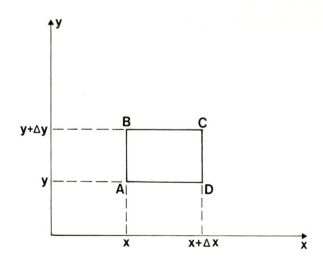

Assumption (b) applied to the rectangle shown tells us that the flow J_{AB} to the right across side AB is given by

$$J_{AB} \simeq -k\,\Delta y\,\frac{\partial p}{\partial x}\,(x, y + \tfrac{1}{2}\Delta y)$$

where p is the pressure function. Similarly

$$J_{AD} \simeq -k\,\Delta x\,\frac{\partial p}{\partial y}\,(x + \tfrac{1}{2}\Delta x, y)$$

$$J_{BC} \simeq -k\,\Delta x\,\frac{\partial p}{\partial y}\,(x + \tfrac{1}{2}\Delta x, y + \Delta y)$$

$$J_{CD} \simeq -k\,\Delta y\,\frac{\partial p}{\partial x}\,(x + \Delta x, y + \tfrac{1}{2}\Delta y).$$

Assumption (c) gives

$$0 = J_{AB} + J_{AD} - J_{BC} - J_{CD}.$$

By the same type of manipulation as in the last few lines of the text (but with p replacing u) we deduce that p satisfies *Laplace's equation*.

32.1.3   Summary of Section 32.1

The main new idea of mathematical modelling introduced here is that of a conservation law. This is a law stating that same physical quantity is neither created nor destroyed and can be put in mathematical form by selecting a well-defined region of space and applying to it the rule

$$\text{accumulation} = \text{input} - \text{output} \qquad (1)$$

In formulating the heat conduction equation in one dimension we apply this to the section between x and $x + \Delta x$; in two-dimensions with rectangular coordinates, to a rectangular section of sides Δx and Δy, and so on. We also use Fourier's law (heat flow is proportional to space rate of change of temperature) and the law that the rate of temperature rise of some portion of matter is proportional to the rate at which heat accumulates in it. The differential equation is obtained by dividing (1) by Δx (or $\Delta x \Delta y$) combining with the other two laws, and taking the limit of small Δx (or Δx and Δy).

In this section we defined the terms

conservation laws	(page C7)	★ ★
steady-state situation	(page C9)	★ ★

Technique

Derive the partial differential equation describing the flow of heat in a   ★
body by applying Fourier's law, the specific heat law and the appropriate
conservation law.

32.2 SOLVING THE EQUATIONS

32.2.0 Introduction

In this section we consider some methods for solving boundary-value problems for partial differential equations of the types derived in sub-section 32.1.2. These methods are mainly based on the technique of separation of variables, which we studied in *Unit 23*.

32.2.1 One-dimensional Heat Conduction: Homogeneous End-point Conditions

The term "homogeneous boundary conditions" as applied to ordinary differential equations was introduced in *Unit 25, Boundary-value Problems* (page **K**458). In the context of heat conduction, homogeneous boundary conditions are those where the temperature at some spatial boundary, or the heat flow across it, or a linear combination of the two, is required to be equal to zero. On the other hand, if the temperature is required to have some value other than zero, we have a nonhomogeneous boundary condition. The homogeneous case can be dealt with by a straight-forward application of the method of separation of variables.

One such problem is that which models a rod whose ends are both maintained at temperature 0 and whose temperature distribution at time 0 is a given function f:

solve

$$\frac{\partial^2 u}{\partial x^2} = a^2 \frac{\partial u}{\partial t} \qquad ((x, t) \in [0, L] \times [0, \infty))$$

with

$$u(0, t) = u(L, t) = 0 \qquad (t \in (0, \infty))$$

and

$$u(x, 0) = f(x) \qquad (x \in [0, L]).$$

We have already seen how to solve this problem by separation of variables in sub-section 23.3.3 on *Unit 23*. You may like to revise this solution, or the very similar one for the wave equation given in Section 13-4 of **K**, before proceeding to the reading passage where a slightly more complicated problem of the same type is considered.

READ Section 13-6 from its start on page **K**528 *to* "... solution of the given problem" *in line* -10, *page* **K**530.

Notes

(i) *Equations (13-38), middle line, page* **K**529   Comparing this with line 14 on page **K**514 and our remarks on Exercise 1 of sub-section 32.1.1, we see that if the end of the rod is open to the air, h is positive (and zero if it is insulated).

(ii) *line 13, page* **K**530   This formula for A_n depends on the fact that the eigenfunctions of a Sturm-Liouville system are orthogonal in the appropriate Euclidean space (here $C[0, L]$). See the equation two lines above Equations (12-31) on page **K**484.

(iii) *Equations (13-42), page* **K**530   In writing this we take advantage of the fact that the eigenfunctions of the Sturm-Liouville problem form a basis (since the boundary conditions are unmixed). See Theorem 12-7 on page **K**487.

Exercises

1. Exercise 2, page **K**531.

2. Exercise 5, page **K**531. The end-point conditions should be read as

$$u_x(0, t) = u_x(L, t) = 0 \qquad (t \in (0, \infty)),$$

excluding the instant $t = 0$, since the given boundary condition

$$u(x, 0) = k \sin \frac{\pi x}{L}$$

tells us that

$$u_x(x, 0) = \frac{k\pi}{L} \cos \frac{\pi x}{L}$$

and so

$$u_x(0, 0) = u_x(L, 0) = 0$$

is untrue.

Solutions

1. Separation of variables with $u(x, t) = X(x)T(t)$ gives the pair
    of differential equations (13-40) on page **K529** with

    $$X(0) = X(L) = 0$$

    whence

    $$X(x) = \sin \frac{n\pi x}{L}, \quad \lambda = -\frac{n^2\pi^2}{L^2} \qquad (n = 1, 2, \ldots)$$

    $$T(t) = A_n \, exp \left(-\frac{n^2\pi^2 t}{L^2 a^2} \right)$$

    where n is any non-zero integer and A_n is any real number.

    Since

    $$T(0) = A_n$$

    the initial condition

    $$u(x, 0) = k \sin \frac{\pi x}{L}$$

    can be satisfied by taking $n = 1$ and $A_n = k$, and we obtain

    $$u(x, t) = k \sin \left(\frac{\pi x}{L} \right) \exp \left(-\frac{\pi^2 t}{L^2 a^2} \right).$$

2. The separation of variables method gives the pair of differ-
    ential equations (13-40) with

    $$X'(0) = X'(L) = 0.$$

    The eigenfunctions X are

    $$X(x) = \cos \frac{n\pi x}{L}, \quad \lambda = -\frac{n^2\pi^2}{L^2} \qquad (n = 0, 1, 2, \ldots)$$

    $$T(t) = A_n \exp \left(-\frac{n^2\pi^2 t}{L^2 a^2} \right).$$

    The series solution is therefore

    $$u(x, t) = \sum_{n=0}^{\infty} A_n \cos \frac{n\pi x}{L} \exp \left(-\frac{n^2\pi^2 t}{L^2 a^2} \right).$$

    To satisfy the initial conditions we want

    $$k \sin \frac{\pi x}{L} = \sum_{n=0}^{\infty} A_n \cos \frac{n\pi x}{L} \qquad (x \in [0, L]).$$

By the theory in Sections 9-5 and 9-6 of **K**, which we studied in *Unit 22, Fourier Series*, the coefficients are given by

$$A_n = \frac{2}{L} \int_0^L k \sin \frac{\pi x}{L} \cos \frac{n\pi x}{L} \, dx$$

if $n = 1, 2, \ldots$ and a similar formula without the factor 2 if $n = 0$. Carrying out the integrations and substituting in the above formula for $u(x, t)$, we obtain the answer given on page **K**758 (in which $2m$ is written in place of n, since only even values of n contribute to the sum).

32.2.2 One-dimensional Heat Conduction: Nonhomogeneous End-point Conditions

A typical pair of nonhomogeneous end-point conditions for the one-dimensional heat conduction equation

$$\frac{\partial^2 u}{\partial x^2} = \frac{1}{a^2} \frac{\partial u}{\partial t} \qquad ((x, t) \in [0, L] \times [0, \infty))$$

is

$$\left. \begin{array}{l} u(0, t) = 100 \\ u(L, t) = 0 \end{array} \right\} \qquad (t \in [0, \infty)) \tag{1}$$

describing a rod (or slab) where two ends (or faces) are maintained at different temperatures.

The difference between this type of end-point condition and the homogeneous type can be seen by looking for the steady-state solution. If $u(x, t)$ is independent of time, it has the form

$$u(x, t) = X(x)$$

where X is some function from $[0, L]$ to R, and the heat conduction equation for u then implies

$$X''(x) = 0.$$

Integrating twice, we obtain first $X'(x) = c_1$ and then

$$X(x) = c_1 x + c_2$$

where c_1 and c_2 are arbitrary constants of integration. Choosing these constants to fit the end-point conditions (1) we obtain as the steady-state solution

$$u(x, t) = 100 \left(1 - \frac{x}{L} \right).$$

If we had done a similar calculation for the homogeneous boundary conditions

$$\left. \begin{array}{l} u(0, t) = 0 \\ u(L, t) = 0 \end{array} \right\} \qquad (t \in [0, \infty)) \tag{2}$$

we would have obtained

$$u(x, t) = 0$$

as our steady-state solution. Thus a major distinction between homogeneous and nonhomogeneous boundary conditions is that for the latter the steady-state solution cannot be 0.

The differences between the two cases can also be stated in algebraic terms. For the homogeneous end-point conditions (2) the set of all solutions u of the heat conduction equation (on the given domain $[0, L] \times [0, \infty)$) which also satisfy these conditions is a vector space, and we can therefore proceed by looking for a basis of this vector space; the solutions obtained by separating variables form such a basis. For the nonhomogeneous end-point conditions (1), on the other hand, the set of solutions of the heat conduction equation satisfying them is not a vector space, but a linear manifold (see sub-section 15.1.2 of *Unit 15, Affine Geometry and Convex Cones.*) The verification of this fact is the subject of Exercise 1 of this sub-section. To specify this linear manifold we proceed in the usual way, expressing u as the sum of a particular solution and a general element of the kernel of the associated homogeneous problem. We already have one particular solution, the steady-state solution, and so we can obtain the general solution by writing

$$u(x, t) = 100\left(1 - \frac{x}{L}\right) + v(x, t).$$

It is then easily verified that v satisfies the one-dimensional heat equation

$$\frac{\partial^2 u}{\partial x^2} = a^2 \frac{\partial u}{\partial t}$$

and the homogeneous end-point conditions

$$\left.\begin{array}{l} v(0,\ t) = 0 \\ v(L, t) = 0 \end{array}\right\} \quad (t \in [0,\ \infty))$$

so that the possible functions v do form a vector space and can be found by the method described in the preceding sub-section.

The next reading passage gives an example of this procedure.

READ the rest of Section 13-6, from line -9, *page* **K**530.

Notes

(i)   *line* -2, *page* **K**530   It is not really necessary to use separation of variables for this particular solution; a more direct procedure is to set out directly to find the steady-state solution by writing $u(x, t) = X(x)$.

(ii)   *line 1, page* **K**531   There is no choice about the value of λ; only by taking $\lambda = 0$ is it possible to have T a constant and so satisfy the nonhomogeneous boundary condition at $x = 0$ for all t. As indicated in the preceding note, the resulting solution u_0 is the steady-state solution.

Exercises

1.   Verify that the solutions of the one-dimensional heat conduction equation, each of which also satisfies the boundary conditions

$$\left.\begin{array}{l} u(0,\ t) = 100 \\ u(L, t) = \quad 0 \end{array}\right\} \quad (t \in [0,\ \infty))$$

for some fixed L form a linear manifold.

2.   A slab of thickness L (a section of the wall of a kettle, for example) is brought to a uniform temperature 100, and then at time 0 a block of ice is pushed against the face $x = 0$, reducing the temperature there to 0 for all times $t > 0$. The face $x = L$ is maintained at temperature 100 by boiling water. Find the temperature distribution in the slab at time t. Use the result:

$$\int_0^L x \sin \frac{n\pi x}{L}\, dx = -\frac{L^2}{n\pi} \cos n\pi x.$$

Solutions

1. Let u_1, u_2 be solutions of the one-dimensional heat conduction equation both of which satisfy the conditions

$$\left.\begin{array}{l} u(0,\ t) = \quad 0 \\ u(L,\ t) = 100 \end{array}\right\} \qquad (t \in [0,\ \infty))$$

Then if α_1, α_2 are real numbers such that $\alpha_1 + \alpha_2 = 1$, the function u_3 defined by

$$u_3 = \alpha_1 u_1 + \alpha_2 u_2$$

is a solution of the heat conduction equation (since u_3 is a linear combination of u_1 and u_2) and it also satisfies

$$u_3(0,\ t) = 0$$
$$u_3(L,\ t) = \alpha_1 \times 100 + \alpha_2 \times 100 = 100$$

Thus the solution set specified in the question is closed under the taking of affine combinations and is therefore a linear manifold.

2. The boundary conditions are

$$\left.\begin{array}{l} u(0,\ t) = \quad 0 \\ u(L,\ t) = 100 \end{array}\right\} \qquad (t \in (0,\ \infty))$$

$$u(x, 0) = 100 \qquad (x \in [0,\ L])$$

As explained in this sub-section, we start by writing

$$u(x,\ t) = 100\left(1 - \frac{x}{L}\right) + v(x,\ t)$$

where v satisfies the one-dimensional heat conduction equation and the boundary conditions

$$\left.\begin{array}{l} v(0,\ t) = 0 \\ v(L,\ t) = 0 \end{array}\right\} \qquad (t \in (0,\ \infty))$$

$$v(x, 0) = 100\,\frac{x}{L} \qquad (x \in (0,\ L))$$

The separation of variables calculation is the same as in Exercise 1 of sub-section 32.2.1 and gives

$$v(x,\ t) = \sum_{n=1}^{\infty} A_n \sin\left(\frac{n\pi x}{L}\right) \exp\left(-\frac{n^2\pi^2 t}{L^2 a^2}\right).$$

The initial conditions require

$$100\,\frac{x}{L} = \sum_{n=1}^{\infty} A_n \sin\left(\frac{n\pi x}{L}\right) \qquad (x \in (0,\ L))$$

and so

$$A_n = \frac{2}{L} \int_0^L 100\,\frac{x}{L} \sin\left(\frac{n\pi x}{L}\right) dx$$

$$= \left(\frac{2}{L}\right)\left(\frac{100}{L}\right)\left(-\frac{L^2(-1)^n}{n\pi}\right) = \frac{-200(-1)^n}{n\pi}$$

giving

$$u(x,\ t) = 100\left(1 - \frac{x}{L}\right)$$

$$+ \frac{200}{\pi} \sum_{n=1}^{\infty} \frac{(-1)^{n+1}}{n} \sin\left(\frac{n\pi x}{L}\right) \exp\left(\frac{-n^2\pi^2 t}{L^2 a^2}\right).$$

CHRIST'S COLLEGE
LIBRARY

(This solution has the disconcerting feature that it makes $u(0, 0)$ and $u(L, 0)$ differ by 100, whereas the problem as set requires them to be equal. The difficulty arises because the initial temperature distribution $u(x, 0) = 100$ does not satisfy the given endpoint conditions. This does not affect the validity of the solution for any other pairs (x, t) except $(0, 0)$ and $(L, 0)$.)

Further practice: Exercises 9, 11 on page **K**532.

32.2.3 Laplace's Equation: Rectangular Region

Boundary-value problems for Laplace's equation

$$\frac{\partial^2 u}{\partial x^2} + \frac{\partial^2 u}{\partial y^2} = 0$$

are somewhat more complicated than those for the one-dimensional heat conduction equation. This is partly because the equation involves second-order partial derivatives with respect to both its variables x and y instead of only one, and partly because the boundary conditions, or even the shape of the boundary in the xy-plane, may be complicated. In this sub-section we consider a type of boundary-value problem for Laplace's equation where these difficulties can be overcome by extending the methods we have been using for one-dimensional heat conduction problems.

READ Section 14-2, which starts on page **K**548.

Notes

(i) *Equations 14-9, page* **K**548 This problem is illustrated by Figure 13-14 on page **K**532. Without the condition $u(L, y) = 0$, the boundary conditions would be exactly the same as for heat conduction in a bar with ends at $x = 0$ and $x = L$, both held at zero temperature—a problem with homogeneous endpoint conditions, which we saw how to solve in sub-section 32.2.1. The method of solution here is very nearly identical.

(ii) *line* -14, *page* **K**550 "Fourier series ... $[-L, L]$" means "Fourier sine series expansion on $[0, L]$".

(iii) *Last paragraph of the section, pages* **K**550–1 We find u_3 by the method described in the reading passage, with f_3 written for f. We find u_4 by solving Laplace's equation with the boundary conditions

$$u(0, y) = 0 \qquad u(L, y) = 0$$
$$u(x, M) = f_4(x) \qquad u(x, 0) = 0$$

by a method closely following that in the reading passage until just after Equation (14-12) where we take $Y(0) = 0$ in place of $Y(M) = 0$. The effect of this is to replace $\sinh\left(\frac{n\pi}{L}(M - y)\right)$ by $\sinh\left(\frac{n\pi y}{L}\right)$ throughout the rest of the calculation. We find u_1 and u_2 by the same method as u_3 and u_4 but with x and y, L and M, X and Y interchanged.

Exercises

1. Exercise 1, page **K**551.

   Use the result:

$$\int_0^L 2x(x - L) \sin\frac{n\pi x}{L}\, dx = \begin{cases} -\dfrac{8L^3}{\pi^3 n^3} & \text{if } n \text{ odd} \\[2mm] 0 & \text{if } n \text{ even} \end{cases}$$

2. Exercise 5, page **K**551.

Solutions

1. As explained at the end of the reading passage, the solution is $u_3 + u_4$ where u_3 and u_4 are solutions of Laplace's equation satisfying

   (a) $u_3(0, y) = u_3(L, y) = u_3(x, M) = 0 \quad u_3(x, 0) = 2x(x - L)$

   (b) $u_4(0, y) = u_4(L, y) = u_4(x, 0) = 0 \quad u_4(x, M) = 2x(x - L)$

   Case (a) is just that discussed in **K** and its solution is obtained by substituting the integral given in the question into Equation (14-15) on page **K**550. We obtain

   $$u_3(x, y)$$
   $$= \frac{-16L^2}{\pi^3} \sum_{n=1, 3, 5, \ldots} \frac{\sin (n\pi x/L) \sinh [n\pi(M - y)/L]}{n^3 \sinh (n\pi M/L)}$$

   For case (b) we proceed as **K** does until Equation (14-12) on page **K**549. We now argue as follows: since $Y(0) = 0$, we have $C_n = 0$ (recall that $\cosh 0 = 1$, $\sinh 0 = 0$) so that

   $$Y_n(y) = B_n \sinh (n\pi y/L).$$

   To obtain the general solution we form the series

   $$u_4(x, y) = \sum_{n=1}^{\infty} A_n B_n \sin \frac{n\pi x}{L} \sinh \frac{n\pi y}{L}.$$

   We may write a_n for the product $A_n B_n$ and so the boundary condition on $y = M$ gives

   $$u_4(x, M) = \sum_{n=1}^{\infty} a_n \sin \frac{n\pi x}{L} \sinh \frac{n\pi M}{L} = 2x(x - L).$$

   Using again the integral given in the question, we obtain

   $$a_n \sinh \frac{n\pi M}{L} = \frac{2}{L} \int_0^L 2x(x - L) \sin \frac{n\pi x}{L} \, dx$$
   $$= \begin{cases} \dfrac{-16L^2}{\pi^3 n^3} & \text{if } n \text{ odd} \\ 0 & \text{if } n \text{ even} \end{cases}$$

   and so the series for u_4 is

   $$u_4(x, y)$$
   $$= -\frac{16L^2}{\pi^3} \sum_{n=1, 3, 5, \ldots} \frac{\sin (n\pi x/L) \sinh (n\pi y/L)}{n^3 \sinh (n\pi M/L)}.$$

   The complete solution is

   $$u(x, y) = u_3(x, y) + u_4(x, y)$$
   $$= -\frac{16L^2}{\pi^3} \sum_{n=1, 3, 5, \ldots} \left(\sin \left(\frac{n\pi x}{L} \right) \right.$$
   $$\left. \times \frac{\sinh (n\pi y/L) + \sinh [n\pi(M - y)/L]}{n^3 \sinh (n\pi M/L)} \right).$$

2. The solution is $u_1 + u_2 + u_3 + u_4$, where apart from a factor 2 u_3 and u_4 have already been calculated in solving Exercise 1, and u_1, u_2 are the solutions of Laplace's equation that satisfy

   $$u_1(0, y) = \sin \frac{\pi y}{M}, \quad u_1(x, 0) = u_1(x, M) = u_1(L, y) = 0$$

and

$$u_2(L, y) = \sin\frac{\pi y}{M}, \quad u_2(x, 0) = u_2(x, M) = u_2(0, y) = 0.$$

For u_1 the method of separation of variables gives $u_1 = XY$ with

$$X'' + \lambda X = 0, \quad Y'' - \lambda Y = 0$$

and

$$Y(0) = Y(M) = 0.$$

The eigenfunctions of this two-point boundary value problem for Y are given by

$$Y(y) = A_n \sin\frac{n\pi y}{M} \quad \text{with} \quad n = 1, 2, 3, \ldots.$$

The eigenvalues are

$$\lambda = -\frac{n^2\pi^2}{M^2}$$

and the corresponding solutions of the equation for X are therefore

$$X(x) = B_n \sinh\frac{n\pi x}{M} + C_n \cosh\frac{n\pi x}{M}$$

The boundary conditions on u_1 imply $X(L) = 0$ and so we obtain (as in the derivation of Equation (14-13) on page **K**549)

$$X(x) = \sinh\frac{n\pi}{M}(L - x)$$

and hence

$$X(x)Y(y) = A_n \sinh\frac{n\pi}{M}(L - x)\sin\frac{n\pi y}{M}.$$

A more general solution can be obtained by summing over n, but for our purposes this is not necessary since the fourth boundary condition,

$$u_1(0, y) = \sin\frac{\pi y}{M},$$

can only be satisfied by taking $n = 1$, to give

$$u_1(x, y) = \frac{\sinh\left(\frac{\pi}{M}(L - x)\right)\sin\frac{\pi y}{M}}{\sinh\frac{\pi L}{M}}.$$

A similar calculation for u_2 gives

$$u_2(x, y) = \frac{\sinh\frac{\pi x}{M}\sin\frac{\pi y}{M}}{\sinh\frac{\pi L}{M}}$$

and so the solution to the original problem is

$$u(x, y) = u_1 + u_2 + u_3 + u_4$$

$$= \sin\frac{\pi y}{M}\left(\sinh\frac{\pi x}{M} + \sinh\frac{\pi}{M}(L - x)\right)$$

$$- \frac{8L^2}{\pi^3}\sum_{n=1, 3, 5, \ldots}\left(\sin\left(\frac{n\pi x}{L}\right)\right.$$

$$\times \left.\frac{\sinh(n\pi y/L) + \sinh[n\pi(M - y)/L]}{n^3 \sinh(n\pi M/L)}\right).$$

32.2.4 Laplace's Equation: Circular Region

In sub-section 32.1.2 we stated Laplace's equation in polar coordinates (it is derived in the Appendix), and commented on its usefulness for boundary-value problems involving the solution of Laplace's equation in a region with a circular boundary. The equation is

$$\frac{\partial^2 u}{\partial r^2} + \frac{1}{r}\frac{\partial u}{\partial r} + \frac{1}{r^2}\frac{\partial^2 u}{\partial \theta^2} = 0 \qquad (r \neq 0) \qquad (1)$$

Before considering its solution by separation of variables (this is done in the reading passage), let us consider what happens when the physical region on which $u(r, \theta)$ is to be defined includes the origin, so that the domain of the function u includes $r = 0$. Mathematically, this point is exceptional since the differential equation (1) is not defined for $r = 0$. Physically, however, there is nothing special about this point and so we require $u(r, \theta)$ to behave in the same way at the origin as it does elsewhere in the physical region. In particular we require u to be continuous at $r = 0$. This condition restricts the solutions of the equation when the physical region includes the origin (compare the remark at the end of Solution 2, sub-section 25.2.1 of *Unit 25*). You will see an example of this in the reading passage, where the function u defined by

$$u(r, \theta) = c \ln kr \quad \text{with} \quad c \neq 0,$$

whose domain does not include $r = 0$, is rejected as a solution of the equation.

READ Section 14-3 from page **K553** *to line 8 on page* **K555**.

Notes

(i) *Equation (14-17), page* **K553** An alternative method of solution is the one used in the method of reduction of order, treated in Section 4-6 of **K** and in sub-section 11.2.5 of *Unit 11, Differential Equations III*. We write $z = \dfrac{du}{dr}$ and treat the equation as a first-order equation in z:

$$\frac{dz}{dr} + \frac{z}{r} = 0.$$

so that

$$z = \frac{c}{r}$$

i.e.

$$\frac{du}{dr} = \frac{c}{r}$$

where c is a constant of integration. This is the same as the result of the first integration of the equation after Equation (14-17). The second integration gives $u = c \ln r + c_1$, since $r > 0$, which is equivalent to Equation (14-18).

(ii) *Equation (14-18), page* **K553** The solution (14-18) is only useful in cases where the domain of u does not include the origin, e.g. steady state heat flow in the annular region shown in Figure 14-3 on page **K557** which models heat conduction between the inside and outside of a pipe.

(iii) *line* -1, *page* **K553** "single-valued" means that the mapping

(position in plane) $\longmapsto$ temperature

is one-one or many-one, i.e. it is a function. It follows that u must map (r, θ) and $(r, \theta + 2\pi)$ to the same number, since they both correspond to the same point in the plane. This gives the periodicity condition

$$u(r, \theta) = u(r, \theta + 2\pi)$$

which applies to *any* solution referring to a circular or annular region, not merely to solutions that are independent of r.

(iv) *line 16, page* K554 "periodic boundary conditions" are defined as case 3 on page K478; in the present problem they would be $\theta(0) = \theta(2\pi)$ *and* $\theta'(0) = \theta'(2\pi)$. An equivalent condition is to give Θ the domain R instead of $(0, 2\pi]$ and require it to be periodic with period 2π. The reason for this periodicity is explained in the preceding note.

(v) *line* -8, *page* K554 We have not studied the Euler equation (it is treated in Section 4-8 of K) but all that is needed here is to observe that the solution space is two-dimensional (since the equation is of second order, and normal on $(0, \infty)$) and that a basis for the solution space is given by $\{r^n, r^{-n}\}$ if $n \neq 0$ and $\{1, \ln r\}$ if $n = 0$.

(vi) *line* -6, *page* K554 The functions $r \longmapsto r^{-n}$ and $r \longmapsto \ln r$ cannot have $r = 0$ in their domains and can therefore be used only if the domain of u does not include the origin.

Exercise

Exercise 3, page K557. (The same mathematics also applies to the problem of a long pipe carrying a liquid at one temperature and immersed in a liquid at another.)

Solution

The problem has rotational symmetry; so we expect $u(r, \theta)$ to be independent of θ. The only θ-independent solution of Laplace's equation is the one given in Equation (14-18) on page K553. The boundary conditions then determine the constants of integration c and k as follows:

$$0 = c \ln kR_1$$

$$100 = c \ln kR_2$$

whence

$$k = \frac{1}{R_1} \quad \text{and} \quad c = \frac{100}{[\ln (R_2/R_1)]}.$$

The required temperature distribution is

$$u(r, \theta) = 100 \frac{\ln (r/R_1)}{\ln (R_2/R_1)}.$$

The points where the temperature is $50°$ are given by

$$50 = 100 \frac{\ln (r/R_1)}{\ln (R_2/R_1)}$$

i.e. $\ln (R_2/R_1) = \ln (r/R_1)^2$.

Hence the points where the temperature is $50°$ form the circle (or cylinder) $r = \sqrt{R_1 R_2}$.

32.2.5 Other Methods of Solution (Optional)

The method of separation of variables is only one of the methods used in solving partial differential equations. There are several others, including Laplace transforms, Fourier transforms, and various numerical methods. We have illustrated the use of Fourier transforms in *Unit 31, Fourier Transforms*. In this sub-section we indicate how Laplace transforms and numerical methods can be used to solve the type of problem discussed in this unit.

To illustrate the use of Laplace transforms, consider again Exercise 2 on page K531. (This is Exercise 1 of sub-section 32.2.1.) The problem is to find a function u satisfying the one-dimensional heat conduction equation

$$\frac{\partial^2 u}{\partial x^2} = a^2 \frac{\partial u}{\partial t}$$

and the conditions

$$u(0, t) = u(L, t) = 0 \qquad (t \in [0, \infty))$$

$$u(x, 0) = k \sin \frac{\pi x}{L} \qquad (x \in [0, L)).$$

We can define the Laplace transform of u, a function of two variables by a device similar to that used in *Unit 31, Fourier Transforms*. Let f_x be the function of one variable defined by

$$t \longmapsto u(x, t) \qquad (t \in [0, \infty)).$$

For each x there is a function f_x. The Laplace transform of f_x is

$$\mathfrak{L}[f_x](s) = \int_0^\infty e^{-st} f_x(t) \, dt. \tag{1}$$

Since

$$f_x(t) = u(x, t)$$

we may regard (1) as defining the Laplace transform of u. We can write (1) as

$$\mathfrak{L}[f_x](s) = \mathfrak{L}[u](x, s) = \int_0^\infty e^{-st} u(x, t) \, dt.$$

We can then take the Laplace transforms of both sides of the heat conduction equation, obtaining

$$\mathfrak{L}\left[\frac{\partial^2 u}{\partial x^2}\right] = a^2 \mathfrak{L}\left[\frac{\partial u}{\partial t}\right].$$

If we assume, for the left-hand side, that differentiation under the integral sign is justified, and apply Theorem 5-4 on page **K187** to the right-hand side, we obtain

$$\frac{\partial^2}{\partial x^2} \mathfrak{L}[u](x, s) = a^2(s\mathfrak{L}[u](x, s) - u(x, 0)).$$

This equation can be written, using the initial condition on u, as

$$\left(\frac{\partial^2}{\partial x^2} - a^2 s\right) \mathfrak{L}[u](x, s) = -a^2 k \sin \frac{\pi x}{L}$$

and solved as an ordinary differential equation for each value of s, subject to the boundary conditions

$$\mathfrak{L}[u](0, s) = \mathfrak{L}[u](L, s) = 0.$$

The solution is

$$\mathfrak{L}[u](x, s) = \frac{a^2 k}{a^2 s + \pi^2/L^2} \sin \frac{\pi x}{L}$$

as you can check by substituting into the differential equation. Finally, we invert the Laplace transform using the table on page **K229**, and obtain

$$u(x, t) = k \exp\left(-\pi^2 t/L^2 a^2\right) \sin \frac{\pi x}{L}.$$

25

For this particular problem separation of variables yields the solution more quickly, but for more complicated problems the Laplace transform method often turns out to be a useful source of information about the solution.

To illustrate the numerical solution of partial differential equations let us consider the solution of Laplace's equation

$$\frac{\partial^2 u}{\partial x^2} + \frac{\partial^2 u}{\partial y^2} = 0$$

on the triangular domain shown below subject to the conditions

$$u(x, 0) = x^2,$$

$$u(0, y) = 0,$$

$$u(x, 1 - x) = x.$$

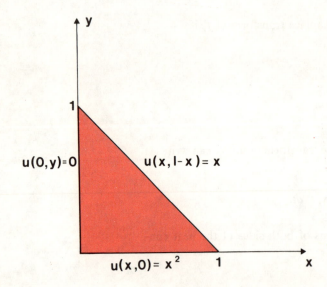

Separation of variables will not work here because there is no suitable co-ordinate system (x_1, x_2) for which the boundaries can be described as $x_i =$ constant. For a numerical solution we choose a step size h and look for approximations to $u(x, y)$ at the points where x and y are integer multiples of h.

The diagram shows these points for the case $h = \frac{1}{4}$, together with those values of u (in red) that are determined directly by the boundary conditions.

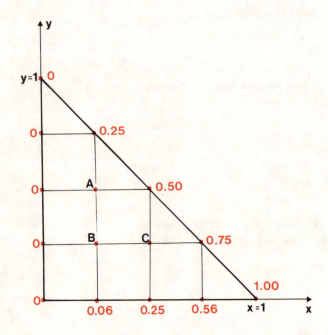

The unknown values u_A, u_B, u_C, where $u_A = u(\frac{1}{4}, \frac{1}{2})$, etc., can now be calculated approximately using an approximate formula for the derivatives in Laplace's equation. This approximation is analogous to the approximation for the second derivative of a function in $C^2(R)$

$$f''(x) = \frac{f(x+h) - 2f(x) + f(x-h)}{h^2} + O(h^2)$$

which we used in the optional sub-section 21.4.2 of *Unit 21, Numerical Solution of Differential Equations*. It is

$$\frac{\partial^2 u}{\partial x^2}(x, y) = \frac{u(x+h, y) - 2u(x, y) + u(x-h, y)}{h^2} + O(h^2)$$

$$\frac{\partial^2 u}{\partial y^2}(x, y) = \frac{u(x, y+h) - 2u(x, y) + u(x, y-h)}{h^2} + O(h^2)$$

so that Laplace's equation becomes (neglecting the terms of order h^2)

$$u(x+h, y) + u(x-h, y) + u(x, y+h) + u(x, y-h) - 4u(x, y) \simeq 0$$

In other words, the value of u at each mesh point of a network like the one shown in the diagram is approximately the average of its values at the four neighbouring mesh points.

Applied to the grid shown above, the approximation requires

(at A) $0.50 + 0 + 0.25 + u_B - 4u_A = 0,$

(at B) $u_C + 0 + u_A + 0.06 - 4u_B = 0,$

(at C) $0.75 + u_B + 0.50 + 0.25 - 4u_C = 0.$

By solving this system of equations (using, for example, Gauss elimination with interchanges) we obtain

$$u_A = 0.23, \quad u_B = 0.18, \quad u_C = 0.42.$$

The accuracy of an approximate solution obtained by this method can be improved by going to a smaller value of h, say half the previous one, and the error estimated by comparing solutions with different values of h.

32.2.6   Summary of Section 32.2

For homogeneous boundary conditions (i.e. boundary conditions compatible with a zero solution) the methods described in *Unit 23* were used. For the heat equation with nonhomogeneous boundary conditions we look for a solution in the form

$u = u_s + v$

where u_s is the steady-state solution, i.e. one that does not depend on the time variable t, and v is a solution which satisfies a *homogeneous* boundary condition. For Laplace's equation in a rectangular region we look for a solution of the form

$u = u_1 + u_2 + u_3 + u_4$

where u_i satisfies the given boundary condition on the ith side of the rectangle and is zero on the other three sides. For Laplace's equation in a region with a circular boundary we use polar coordinates and apply the boundary condition that u must be a periodic function of the angle θ.

No new terms were defined in this section.

Techniques

1. Use the separation of variables technique to solve the one-dimensional heat conduction equation with homogeneous and nonhomogeneous boundary conditions.   ⋆ ⋆

2. Solve Laplace's equation for rectangular and circular regions by applying the separation of variables technique.   ⋆ ⋆

CHRIST'S COLLEGE
LIBRARY

32.3 PARTIAL DIFFERENTIAL EQUATIONS OF SECOND ORDER

32.3.0 Introduction

We have studied a number of different second-order partial differential equations, but the problems and results were somewhat different in different cases. In this final section we look at partial differential equations of second order in a more general way, to understand better the reasons for these differences. We begin by comparing the main equations we have been studying, and the boundary conditions we have used for each.

32.3.1 Well-posed and Ill-posed Boundary-value Problems

Written in a standardized notation to make them easier to compare and with $a = 1$, our three main partial differential equations are:

the wave equation

$$\frac{\partial^2 u}{\partial x_1^2} - \frac{\partial^2 u}{\partial x_2^2} = 0$$

the heat conduction equation

$$\frac{\partial^2 u}{\partial x_1^2} - \frac{\partial u}{\partial x_2} = 0$$

Laplace's equation (*steady heat conduction*)

$$\frac{\partial^2 u}{\partial x_1^2} + \frac{\partial^2 u}{\partial x_2^2} = 0.$$

(We shall bring Laplace's equation in polar co-ordinates into the story in the next sub-section.)

Superficially the equations look quite similar, but in fact they are very different; in particular the type of boundary condition used is different for each equation. Typical boundary conditions are illustrated below.

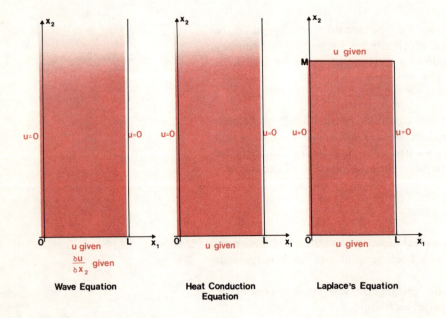

Wave Equation   Heat Conduction Equation   Laplace's Equation

For the wave equation we have two conditions when $x_2 = 0$; for the heat equation, one; and for the Laplace equation we have one condition again but this time there is a second on the line $x_2 = M$. When we approached the equations from physical problems, these differences arose naturally from the differences in the problems; but the object of the present section is to relate these differences to the natures of the equations themselves and not to the physical situations that they model.

We can understand the differences if we look at the solutions obtained by separating the variables, writing

$$u(x_1, x_2) = X_1(x_1) X_2(x_2).$$

The equation for X_1 is, in each of the three cases,

$$X_1'' + \lambda X_1 = 0$$

with solutions

$$X_1(x_1) = A \sin\left(\frac{n\pi x_1}{L}\right), \text{ where } A \text{ is arbitrary,}$$

and eigenvalues

$$\lambda = \frac{n^2 \pi^2}{L^2} \quad (n = 1, 2, \dots).$$

corresponding to the operator $-D^2$.

The equation for X_2 is different in each case. For the wave equation it is

$$X_2'' + \lambda X_2 = 0 \tag{1}$$

For the heat conduction equation it is

$$X_2' + \lambda X_2 = 0 \tag{2}$$

and for Laplace's equation it is

$$X_2'' - \lambda X_2 = 0 \tag{3}$$

For Equations (1) and (3) we need two boundary conditions to determine a solution, and for Equation (2) we need one: but just what type of boundary conditions are most suitable? Remember that, because of the condition $\lambda = n^2 \pi^2 / L^2$, we must be prepared for very large positive values of λ.

We have a clue to the answer from our study of ill-conditioning in *Unit 7, Numerical Solution of Simultaneous Algebraic Equations*, and *Unit 21, Numerical Solution of Differential Equations*. We saw there that if a recurrence relation or differential equation has an increasing exponential solution in its kernel, then any initial-value problem for it will be (absolutely) ill-conditioned; but if the solutions are bounded or decreasing as x increases, then the initial-value problem will be (absolutely) well-conditioned. Applying this criterion to Equations (1), (2), (3) above, we see that initial-value problems for Equations (1) and (2) should be well-conditioned, for all λ, since the kernel of the operator $D^2 + \lambda$ associated with Equation (1) is spanned by the bounded functions $x \longmapsto \sin\sqrt{\lambda}x$, and $x \longmapsto \cos\sqrt{\lambda}x$, and the kernel associated with Equation (2) is spanned by the decreasing function $x \longmapsto \exp(-\lambda x)$. On the other hand, for Equation (3) the kernel is spanned by $x \longmapsto \exp(\sqrt{\lambda}x)$ and $x \longmapsto \exp(-\sqrt{\lambda}x)$, and since one of these is an increasing exponential we would expect an initial-value problem in this case to be ill-conditioned. Arguing back from Equations (1), (2), (3) to the partial differential equations they came from, we may expect, therefore, to find that initial-value problems are well-behaved for the wave and heat conduction equations, but not for Laplace's equation. And it is indeed for the wave and heat conduction equation that we have been using initial-value problems (see the illustrations above), with two initial conditions for the wave equation because its order in x_2 is two and one for the heat equation, whose order in x_2 is only one.

This argument can be carried further. We saw in sub-section 7.3.2 of *Unit 7* how to formulate well-conditioned boundary-value problems for second-order recurrence relations with one increasing and one decreasing exponential solution in the kernel. These were the two-point boundary-value problems. Just the same considerations apply to the second-order differential equation (3)

$$X_2'' - \lambda X_2 = 0.$$

If $X_2(0)$ and $X_2(M)$ are given, we have a two-point boundary-value problem which is well-conditioned. Arguing back to the partial differential equation, we may expect to find that a two-point boundary-value problem for Laplace's equation, with the value of $u(x_1, x_2)$ specified for $x_2 = 0$ and for $x_2 = M$, to be well-behaved; and it is indeed for Laplace's equation that we have been using this type of boundary condition.

The conclusions we have drawn from our knowledge of ill-conditioned and well-conditioned problems are confirmed by the general theory of partial differential equations. This theory includes results which do for partial differential equations what the existence and uniqueness theorem (Theorem 3-2, page K104) does for ordinary differential equations. In particular it tells us that the three boundary-value problems illustrated in the above diagram have unique solutions, provided the given boundary values satisfy suitable conditions (e.g. continuity). These boundary-value problems are said to be *well-posed*. On the other hand if we try to specify the boundary values in one of the ways that led to ill-conditioning in our discussion—for example by giving values of $u(x_1, 0)$ and $\dfrac{\partial u}{\partial x_2}(x_1, 0)$ on the boundary $x_2 = 0$ for Laplace's equation instead of for the wave equation—then it is quite possible that the boundary-value problem has no solution at all. Such problems are said to be *ill-posed*.

Exercises

1. It is proposed to calculate $u(x_1, x_2)$ for $0 \leqslant x_1 \leqslant 1$, $0 \leqslant x_2 \leqslant 1$, given that u satisfies the equation

    $$\frac{\partial^2 u}{\partial x_1^2} + \frac{\partial u}{\partial x_2} = 0$$

    with

    $$u(0, x_2) = u(1, x_2) = 0 \qquad (x_2 \in [0, 1])$$

    and

    $$u(x_1, 0) = f(x_1) \qquad (x_1 \in [0, 1])$$

    where f is a given continuous function, with $f(0) = f(1) = 0$. Using an argument based on separation of variables similar to that given in the text, would you expect this problem to be well- or ill-posed?

2. Consider the following boundary-value problem:

    $$\frac{\partial^2 u}{\partial x_1^2} = \frac{\partial^2 u}{\partial x_2^2}$$

    with

    $$u(0, x_2) = u(1, x_2) = 0 \qquad (x_2 \in [0, 2])$$
    $$\left. \begin{array}{l} u(x_1, 0) = f(x_1) \\ u(x_1, 2) = g(x_1) \end{array} \right\} \qquad (x_1 \in [0, 1])$$

    where f and g are given continuous functions, with $f(0) = f(1) = g(0) = g(1) = 0$.

30

(i) Without doing any calculation, and relying on the results stated above, would you expect this problem to be well- or ill-posed?

(ii) Assuming that $f \neq g$, try to carry through the solution of this problem by separation of variables.

Solutions

1. Separation of variables with

$$u(x_1, x_2) = X_1(x_1)X_2(x_2)$$

gives

$$X_1'' + \lambda X_1 = 0 \quad \text{with} \quad X_1(0) = X_1(1) = 0$$

so that

$$X_1(x_1) = A \sin n\pi x_1 \qquad (x_1 \in [0, 1])$$

and

$$\lambda = n^2\pi^2 \qquad (n = 1, 2, 3, \ldots).$$

The equation for X_2 is

$$X_2' - \lambda X_2 = 0$$

so that

$$X_2(x_2) = c \exp (n^2\pi^2 x_2).$$

Since this solution is an *increasing* exponential, the initial-value problem for X_2 is ill-conditioned, so we expect that problem of finding $u(x_1, x_2)$ from its initial values at $x_2 = 0$ to be ill-posed. There is, in general, no solution.

2. (i) For the wave equation we should have the initial values of u and its derivative, not the initial and final values of u. The problem is ill-posed.

    (ii) Separation of variables as in Solution 1 gives the same solution for X_1, with $\lambda = n^2\pi^2$, and the equation for X_2 is

$$X_2'' + \lambda X_2 = 0$$

giving

$$X_2(x_2) = c_1 \cos n\pi x_2 + c_2 \sin n\pi x_2$$

which implies $X_2(0) = X_2(2)$. Thus every solution obtained by separation of variables has

$$u(x_1, 0) = u(x_1, 2)$$

for all values of x. The same will be true of any linear combination of such solutions. Thus a solution is possible only if f and g are the same function, contrary to the assumption that $f \neq g$.

32.3.2 The Classification of Partial Differential Equations

In our discussion of boundary conditions in the preceding sub-section we did not give separate consideration to Laplace's equation for polar coordinates. Since this equation is merely the Cartesian form of Laplace's equation transformed to a new coordinate system, we would expect this equation to have the same essential properties as the Cartesian form: that is, we would expect to find that a boundary-value problem with the value of u specified on a closed boundary is well-posed and that a problem with both u and $\dfrac{\partial u}{\partial x}$ specified on some part of the boundary, is not well-posed. This turns out to be the case. In fact we can go further and specify a general class of equations for which giving the value of $u(x_1, x_2)$ for (x_1, x_2) on the boundary gives a well-posed problem. These equations are called *elliptic* differential equations.

The general linear homogeneous partial differential equation of second order is

$$p \frac{\partial^2 u}{\partial x_1^2} + 2q \frac{\partial^2 u}{\partial x_1 \, \partial x_2} + r \frac{\partial^2 u}{\partial x_2^2} + s \frac{\partial u}{\partial x_1} + t \frac{\partial u}{\partial x_2} + wu = 0$$

where p, q, r, s, t, w are given functions with the same domain as u. An equation of this type is said to be *elliptic* if the quadratic form

$$(\xi, \eta) \longmapsto p(x_1, x_2)\xi^2 + 2q(x_1, x_2)\xi\eta$$
$$+ r(x_1, x_2)\eta^2 \qquad ((\xi, \eta) \in R^2)$$

is positive definite* for all x_1, x_2 in the domain of u (because the graph of such a quadratic form is an ellipse). For example, with Laplace's equation

$$\frac{\partial^2 u}{\partial x_1^2} + \frac{\partial^2 u}{\partial x_2^2} = 0$$

the quadratic form is

$$(\xi, \eta) \longmapsto \xi^2 + \eta^2$$

and for Laplace's equation in polar coordinates

$$\frac{\partial^2 u}{\partial x_1^2} + \frac{1}{x_1} \frac{\partial u}{\partial x_1} + \frac{1}{x_1^2} \frac{\partial^2 u}{\partial x_2^2} = 0 \qquad (x_1 \neq 0)$$

it is

$$(\xi, \eta) \longmapsto \xi^2 + \frac{1}{x_1^2} \eta^2$$

Thus both forms of Laplace's equation are elliptic. The significance of the quadratic form is this: by changing to a new coordinate system (e.g. from rectangular to polar coordinates) we can alter the form of the equation, but all that happens to the quadratic form is a congruence transformation (see sub-section 14.1.2 of *Unit 14, Bilinear and Quadratic Forms*) and so its positive definite character cannot alter. The proof of this fact is laborious, so we omit it.

The simplest way of recognizing an elliptic partial differential equation is to consider the function $pr - q^2$ (which is the determinant of the matrix of the quadratic form). If $pr - q^2$ takes positive values for all (x_1, x_2) in the domain of u, then the equation is elliptic.

Similar considerations apply to the wave and heat equations. For the wave equation, the quadratic form is

$$(\xi, \eta) \longmapsto \xi^2 - \eta^2$$

with rank 2 and signature 0. Any partial differential equation whose quadratic form has rank 2 and signature 0 is said to be *hyperbolic* (because the graph of such a quadratic form is a hyperbola). Hyperbolic differential equations may be recognized by using the fact that if $pr - q^2$ takes negative values for all (x_1, x_2) in the domain of u, then the equation is hyperbolic.

* See page N 168.

For the heat equation, the quadratic form is

$$(\xi, \eta) \longrightarrow \xi^2$$

with rank 1 and signature 1. Any partial differential equation whose quadratic form has rank 1 and signature 1 is called *parabolic* (because the graph of such a quadratic form is a parabola). Parabolic differential equations may be recognized by using the fact that if $pr - q^2$ has the value zero for all (x_1, x_2) in the domain of u, then the equation is parabolic.

For hyperbolic and parabolic equations the problem of finding u inside a region when its values are given on the boundary is not well-posed. There is a general theory for these equations which specifies types of boundary-value problems for these equations which are well-posed, but it is beyond the scope of this course.

Exercises

1. Classify the following equations as elliptic, hyperbolic or parabolic.

    (i) $\dfrac{\partial^2 u}{\partial x_1^2} + 4 \dfrac{\partial^2 u}{\partial x_1 \, \partial x_2} + 5 \dfrac{\partial^2 u}{\partial x_2^2} + \dfrac{\partial u}{\partial x_1} + x_1^2 u = 0.$

    (ii) $\dfrac{\partial^2 u}{\partial x_1 \, \partial x_2} = 0.$

    (iii) $\dfrac{\partial^2 u}{\partial x_1^2} + 4 \dfrac{\partial^2 u}{\partial x_1 \, \partial x_2} + 4 \dfrac{\partial^2 u}{\partial x_2^2} + 3 \dfrac{\partial u}{\partial x_2} = 0.$

2. This question requires a familiarity with the chain rule for partial differentiation. If you are not familiar with this rule, treat the exercise as an example or omit it completely; you will not be examined on it.

    (i) Given that a function u

    $$u: (x_1, x_2) \longmapsto u(x_1, x_2) \qquad ((x_1, x_2) \in R^2)$$

    satisfies

    $$p \frac{\partial^2 u}{\partial x_1^2} + 2q \frac{\partial^2 u}{\partial x_1 \, \partial x_2} + r \frac{\partial^2 u}{\partial x_2^2} = 0$$

    where p, q, r are real numbers, what equation must the function v, defined by

    $$u(x_1, x_2) = v(ax_1 + bx_2, cx_1 + dx_2) \qquad (x_1, x_2 \in R)$$

    satisfy? Note that the two functions u, v represent the same physical quantity described in two different coordinate systems.

    (*Hint:*

    $$\frac{\partial u}{\partial x_1}(x_1, x_2) = (aD_1 + cD_2)v(ax_1 + bx_2, cx_1 + dx_2)$$

    where

    $$D_1 v(y_1, y_2) = \frac{\partial v}{\partial y_1}(y_1, y_2) \qquad ((y_1, y_2) \in R^2).)$$

    (ii) Show that if the matrix $P = \begin{bmatrix} a & c \\ b & d \end{bmatrix}$ is non-singular, the quadratic forms associated with the equations for u and v are congruent (i.e. their matrices U and V are related by $V = P^T U P$), and hence that the equations for u and v are both of the same type (both elliptic, or both parabolic, or both hyperbolic).

Solutions

1. (i) Elliptic: since the quadratic form

$$(\xi, \eta) \longmapsto \xi^2 + 4\xi\eta + 5\eta^2 = (\xi + 2\eta)^2 + \eta^2$$

has rank 2, signature 2.

(ii) Hyperbolic: since the quadratic form

$$(\xi, \eta) \longmapsto \xi\eta = \tfrac{1}{4}(\xi + \eta)^2 - \tfrac{1}{4}(\xi - \eta)^2$$

has rank 2, signature 0.

(iii) Parabolic: since the quadratic form

$$(\xi, \eta) \longmapsto \xi^2 + 4\xi\eta + 4\eta^2 = (\xi + 2\eta)^2$$

has rank 1, signature 1.

2. (i) Since $u(x_1, x_2) = v(ax_1 + bx_2, cx_1 + dx_2)$, we find

$$\frac{\partial u}{\partial x_1}(x_1, x_2) = (aD_1 + cD_2)v(\ldots, \ldots)$$

(as given)

$$\frac{\partial u}{\partial x_2}(x_1, x_2) = (bD_1 + dD_2)v(\ldots, \ldots)$$

$$\frac{\partial^2 u}{\partial x_1^2}(x_1, x_2) = (aD_1 + cD_2)^2 v(\ldots, \ldots)$$

$$\frac{\partial^2 u}{\partial x_1 \, \partial x_2}(x_1, x_2)$$

$$= (aD_1 + cD_2)(bD_1 + dD_2)v(\ldots, \ldots)$$

$$\frac{\partial^2 u}{\partial x_2^2}(x_1, x_2) = (bD_1 + dD_2)^2 v(\ldots, \ldots)$$

and hence

$$\left(p\frac{\partial^2 u}{\partial x_1^2} + 2q\frac{\partial^2 u}{\partial x_1 \, \partial x_2} + r\frac{\partial^2 u}{\partial x_2^2}\right)(x_1, x_2)$$

$$= (p(aD_1 + cD_2)^2 + 2q(aD_1 + cD_2)$$

$$\times (bD_1 + dD_2) + r(bD_1 + dD_2)^2)v(\ldots, \ldots).$$

so that the differential equation for v is

$$((pa^2 + 2qab + rb^2)D_1^2 + 2(pac + q(ad + cb)$$

$$+ rbd)D_1 D_2 + (pc^2 + qcd + rd^2)D_2^2)v = 0.$$

(ii) The matrix of the quadratic form associated with this differential equation is

$$V = \begin{bmatrix} pa^2 + 2qab + rb^2 & pac + q(ad + cb) + rbd \\ pac + q(ad + cb) + rbd & pc^2 + 2qcd + rd^2 \end{bmatrix}$$

and it is easily verified that

$$V = P^T U P$$

where

$$P = \begin{bmatrix} a & c \\ b & d \end{bmatrix}$$

and U, the matrix of the quadratic form associated with the original differential equation is

$$U = \begin{bmatrix} p & q \\ q & r \end{bmatrix}.$$

The matrices U and V therefore have the same signature, and the equations are of the same type.

32.3.3   Summary of Section 32.3

A boundary-value problem for a partial differential equation is said to be *well-posed* if it has a unique solution; *ill-posed* if it has no solution or more than one. The problem of solving Laplace's equation inside a region given values for u on the boundary of the region is well-posed. The problem of solving the heat equation with $u(0, t)$ and $u(L, t)$ given for all $t > 0$ and $u(x, 0)$ given for all $x \in [0, L]$ is well-posed, and so is that of solving the wave equation with $u(0, t)$ and $u(L, t)$ given for all $t > 0$, and $u(x, 0)$ and $\frac{\partial u}{\partial t}(x, 0)$ given for all $x \in [0, L]$.

A partial differential equation of the form

$$p\frac{\partial^2 u}{\partial x_1^2} + 2q\frac{\partial^2 u}{\partial x_1 \partial x_2} + r\frac{\partial^2 u}{\partial x_2^2} + s\frac{\partial u}{\partial x_1} + t\frac{\partial u}{\partial x_2} + wu = 0$$

(with $u: (x_1, x_2) \longmapsto u(x_1, x_2)$) is said to be *elliptic* if $pr > q^2$, *parabolic* if $pr = q^2$ and *hyperbolic* if $pr < q^2$. Boundary-value problems for elliptic equations are analogous to those for Laplace's equation; for parabolic equations, analogous to those for the heat equation; and for hyperbolic equations, analogous to those for the wave equation.

In this section we defined the terms

well-posed boundary-value problem	(page C30)	⋆
ill-posed boundary-value problem	(page C30)	⋆
elliptic differential equation	(page C32)	⋆
hyperbolic differential equation	(page C32)	⋆
parabolic differential equation	(page C33)	⋆

Techniques

1. Determine whether a boundary value problem is well- or ill-posed.   ⋆
2. Classify a given partial differential equation as elliptic, hyperbolic or parabolic.   ⋆

32.4 SUMMARY OF THE UNIT

We first of all constructed the partial differential equation which would give the heat distribution in a body as a function of position and time. The one dimensional case resulted in an equation of the form

$$\frac{\partial^2 u}{\partial x^2} = a^2 \frac{\partial u}{\partial t}$$

called the one-dimensional heat conduction equation. We then looked at a steady-state situation in two dimensions and derived the equation

$$\frac{\partial^2 u}{\partial x^2} + \frac{\partial^2 u}{\partial y^2} = 0$$

which is called Laplace's equation in two dimensions. However, we found that for problems defined on regions with circular boundaries we have to re-express Laplace's equation in terms of polar co-ordinates

$$\frac{\partial^2 u}{\partial r^2} + \frac{1}{r} \frac{\partial u}{\partial r} + \frac{1}{r^2} \frac{\partial^2 u}{\partial \theta^2} = 0 \qquad (r \neq 0).$$

All these equations are soluble using the separation of variables technique and examples of all three equations were given to show how this is done. We also showed (in an optional sub-section) how Laplace transforms and numerical methods can also be used to solve these equations.

In the last section we looked at general second-order partial differential equations in a general way to see how and why the different boundary conditions imposed produce different solutions. This led on to the question of whether a problem is ill- or well-posed which is dependent on how the boundary conditions are specified.

Definitions

conservation laws	(page C7)	★ ★
steady-state situation	(page C9)	★ ★
well-posed boundary-value problem	(page C30)	★
ill-posed boundary-value problem	(page C30)	★
elliptic differential equation	(page C32)	★
hyperbolic differential equation	(page C32)	★
parabolic differential equation	(page C33)	★

Techniques

1. Derive the partial differential equation describing the flow of heat in a body by applying Fourier's law, the specific heat law and the appropriate conservation law. ★
2. Use the separation of variables technique to solve the one-dimensional heat conduction equation with homogeneous and nonhomogeneous boundary conditions. ★ ★
3. Solve Laplace's equation for rectangular and circular regions by applying the separation of variables technique. ★ ★
4. Determine whether a boundary-value problem is well- or ill-posed. ★
5. Classify a given partial differential equation as elliptic, hyperbolic or parabolic. ★

32.5 SELF-ASSESSMENT

Self-assessment Test

This Self-assessment Test is designed to help you test your understanding of the unit. It can also be used, together with the summary of the unit, for revision. The answers to these questions will be found on the next non-facing page. We suggest that you complete the whole test before looking at the answers.

1. In the derivation of the one-dimensional heat conduction equation in sub-section 32.1.1 we assumed that the thermal conductivity k is constant. What would the equation be for a rod composed of a nonhomogeneous material whose conductivity varied along its length?

   Is it still possible to solve the problem by separation of variables and an eigenfunction expansion?

2. A disk of radius b, initially at temperature 0 throughout, is brought into contact at time $t = 0$ with a heat source, which maintains the temperature of the rim at 1000. Find the temperature at points in the disk for time $t > 0$. *Hint:* The solution of

$$rR'' + R' + \lambda^2 rR = 0$$

   which is bounded for $r \leqslant b$ is $R(r) = J_0(\lambda r)$. Also

$$\sum_{n=1}^{\infty} \frac{J_0(\lambda_n r)}{\lambda_n b J_1(\lambda_n b)} = \tfrac{1}{2} \quad \text{for} \quad r \in (0, b)$$

   where $\lambda_1, \lambda_2, \lambda_3, \ldots$ are the solutions of $J_0(\lambda b) = 0$. The functions J_0 and J_1 are known as zeroth and first-order Bessel functions.

3. Laplace's equation in three dimensions can be written

$$\left[D_r(r^2 D_r) + \frac{1}{\sin \phi} D_\phi (\sin \phi D_\phi) + \frac{1}{\sin^2 \phi} D_{\theta\theta} \right] u(r, \phi, \theta) = 0.$$

   Find a series solution by the method of separation of variables, assuming that u is independent of θ. *Hint:* let $\cos \phi = \mu$.

4. Exercise 2, page K557.

5. Exercise 1, page K531.

6. Exercise 10, page K532.

Solutions to Self-assessment Test

1. The point at which the derivation differs from that on page K512 is line -1. Instead, we should write

    $$\Delta H = -\left[-k\left.\frac{\partial u}{\partial x}\right|_{x+\Delta x} + k\left.\frac{\partial u}{\partial x}\right|_{x}\right]$$

    where k is a function with domain R and codomain R^+. It then follows that

    $$\frac{1}{\Delta x}\left[k\left.\frac{\partial u}{\partial x}\right|_{x+\Delta x} - k\left.\frac{\partial u}{\partial x}\right|_{x}\right] = c\rho\,\frac{\partial u}{\partial t}.$$

    Taking the limit for small Δx, we obtain

    $$\frac{\partial}{\partial x}\left(k(x)\,\frac{\partial u}{\partial x}(x, t)\right) = c\rho\,\frac{\partial u}{\partial t}(x, t).$$

    This is again a second-order linear partial differential equation.

    If $u(x, t) = X(x)T(t)$, then X satisfies

    $$(kX')' = \lambda X$$

    which is clearly in self-adjoint form. With suitable boundary conditions this is a Sturm-Liouville problem and its solutions form a complete set of eigenfunctions. For example, if $k: x \longmapsto 1 - x^2$ and the rod occupies the region $x \in [-1, 1]$, X is a Legendre polynomial.

2. The heat conduction equation is best written in polar coordinates

    $$u_{rr} + \frac{1}{r}\,u_r + \frac{1}{r^2}\,u_{\theta\theta} = a^2 u_t \qquad (r \neq 0).$$

    Since the initial temperature distribution is independent of θ, u will be independent of θ at all t, i.e. $u_{\theta\theta} = 0$. We must solve

    $$u_{rr} + \frac{1}{r}\,u_r = a^2 u_t \qquad ((r, t) \in (0, b) \times R^+).$$

    Let us write $u(r, t) = R(r)T(t)$. Then

    $$\left[R''(r) + \frac{1}{r}\,R'(r)\right]T(t) = a^2 R(r)T'(t).$$

    Thus

    $$rR''(r) + R'(r) + \lambda^2 rR(r) = 0 \qquad (1)$$

    and

    $$T' + \frac{\lambda^2}{a^2}\,T = 0$$

    The boundary condition is $u(b, t) = 1000$ which is nonhomogeneous. We first obtain a particular steady-state solution with this boundary condition.

    $$u^{(s)}(r, t) = 1000.$$

    To this we add the general solution with the homogeneous boundary condition $u(b, t) = 0$, i.e. $R(b) = 0$. We must solve (1) subject to this condition. Now the general solution of (1) which is bounded for the disk $(r < R)$ is $J_0(\lambda r)$; in general $J_0(\lambda b) \neq 0$. So we have an eigenvalue problem—only those values λ_n which satisfy $J_0(\lambda_n b) = 0$ are admissible. The corresponding solutions for T are

    $$t \longmapsto e^{-(\lambda_n/a)^2 t}$$

Thus the solution of the homogeneous problem is

$$u^{(h)}(r, t) = \sum_{n=1}^{\infty} A_n J_0(\lambda_n r)\, e^{-(\lambda_n/a)^2 t} \qquad ((r, t) \in (0, b) \times R^+)$$

The solution of the nonhomogeneous problem is

$$u = u^{(s)} + u^{(h)}$$

and we require $u(r, 0) = 0$, $r \in (0, b)$, i.e.

$$1000 + \sum_{n=1}^{\infty} A_n J_0(\lambda_n r) = 0.$$

Now

$$\sum_{n=1}^{\infty} \frac{J_0(\lambda_n r)}{\lambda_n b J_1(\lambda_n b)} = \tfrac{1}{2} \quad \text{for} \quad r \in (0, b)$$

therefore

$$A_n = \frac{-2000}{\lambda_n b J_1(\lambda_n b)}$$

and

$$u(r, t) = 1000 \left[1 - \sum_{n=1}^{\infty} \frac{2 J_0(\lambda_n r)}{\lambda_n b J_1(\lambda_n b)}\, e^{-(\lambda_n/a)^2 t} \right].$$

3. If you are stuck, then, before reading on, consult sub-section 25.2.3 of *Unit 25*.

The general solution is

$$\sum_{n=0}^{\infty} \{ A_n r^n + B_n r^{-(n+1)} \} P_n(\cos \phi).$$

For the method, see page K559.

4. Let

$$R_1 : r \longmapsto r^n \qquad (r \in R^+)$$

$$R_2 : r \longmapsto r^{-n} \qquad (r \in R^+).$$

Then

$$R_1' : r \longmapsto n r^{n-1}$$

$$R_1'' : r \longmapsto n(n-1) r^{n-2}.$$

and

$$r^2 R_1'' + r R_1' - n^2 R_1 : r \longmapsto n(n-1) r^n + n r^n - n^2 r^n = 0.$$

Also

$$R_2' : r \longmapsto -n r^{-n-1}$$

$$R_2'' : r \longmapsto n(n+1) r^{-n-2}$$

and

$$r^2 R_2'' + r R_2' - n^2 R_2 : r \longmapsto n(n+1) r^{-n} - n r^{-n} - n^2 r^{-n} = 0.$$

5. Separation of variables with $u(x, t) = X(x) T(t)$ gives the pair of differential equations

$$X'' - \lambda X = 0$$

$$T' - \frac{\lambda}{a^2} T = 0$$

The conditions $X(0) = X(L) = 0$ imply

$$X(x) = \sin \frac{n\pi x}{L}, \qquad \lambda = \frac{-n^2\pi^2}{L^2} \qquad (n = 1, 2, \ldots)$$

$$T(t) = A_n \exp\left(\frac{-n^2\pi^2 t}{L^2 a^2}\right)$$

The series solution is therefore

$$u(x, t) = \sum_{n=1}^{\infty} A_n \sin\left(\frac{n\pi x}{L}\right) \exp\left(\frac{-n^2\pi^2 t}{L^2 a^2}\right).$$

The initial condition gives

$$u(x, 0) = \sum_{n=1}^{\infty} A_n \sin \frac{n\pi x}{L} = \begin{cases} x & 0 < x < \dfrac{L}{2} \\[2mm] L - x & \dfrac{L}{2} \leqslant x < L \end{cases}$$

so that

$$A_n = \frac{2}{L}\left\{\int_0^{L/2} x \sin \frac{n\pi x}{L}\, dx + \int_{L/2}^{L} (L - x) \sin\frac{n\pi x}{L}\, dx\right\}$$

$$= \frac{4L}{n^2\pi^2} \sin \frac{n\pi}{2}$$

$$u(x, t) = \begin{cases} \dfrac{4L}{\pi^2} \displaystyle\sum_{n=1}^{\infty} \dfrac{1}{n^2} \exp\left(\dfrac{-n^2\pi^2 t}{L^2 a^2}\right) \sin\left(\dfrac{n\pi x}{L}\right) & (n\ \text{odd}) \\[4mm] 0 & (n\ \text{even}) \end{cases}$$

6. Since the boundary conditions are nonhomogeneous we first solve the steady-state problem by looking for a solution of the form $u(x, t) = X(x)$. For this u to satisfy the heat conduction equation, X must satisfy

$$X'' = 0$$

and the boundary conditions give

$$X'(0) = X'(L) = k.$$

The differential equation for X gives $X(x) = c_1 x + c_2$ and the boundary conditions give $c_1 = k$. Any value of c_2 will do, so we take $c_2 = 0$ for our particular solution, obtaining

$$u_0(x, t) = kx.$$

The given problem now reduces to

$$u(x, t) = kx + v(x, t)$$

where $v(x, t)$ satisfies the one-dimensional heat conduction equation with

$$\frac{\partial v}{\partial x}(0, t) = \frac{\partial v}{\partial x}(L, t) = 0$$

$$v(x, 0) = -kx.$$

Separation of variables gives

$$v(x, t) = X(x)T(t)$$

with

$$X'' = \lambda X \quad \text{and} \quad X(0) = X(L) = 0$$

$$T' = \lambda T.$$

so that

$$X(x) = \sin\frac{n\pi x}{L}, \quad \lambda = -\frac{n^2\pi^2}{L^2} \qquad (n = 1, 2, \ldots)$$

$$T(t) = A_n \exp\left(-\frac{n^2\pi^2 t}{L^2}\right).$$

The series solution is therefore

$$v(x, t) = \sum_{n=1}^{\infty} A_n \sin\left(\frac{n\pi x}{L}\right) \exp\left(-\frac{n^2\pi^2 t}{L^2}\right).$$

The initial condition gives

$$-kx = v(x, 0) = \sum_{n=1}^{\infty} A_n \sin\frac{n\pi x}{L}$$

so that

$$A_n = \frac{2}{L}\int_0^L (-kx)\sin\frac{n\pi x}{L}\,dx$$

$$= \frac{2kL}{n\pi}(-1)^n.$$

The solution is therefore

$$u(x, t) = kx + \frac{2kL}{\pi}\sum_{n=1}^{\infty}\frac{(-1)^n}{n}\sin\left(\frac{n\pi x}{L}\right)\exp\left(-\frac{n^2\pi^2 t}{L^2}\right).$$

32.6 APPENDIX

Derivation of Laplace's Equation in Polar Co-ordinates

As for the derivation of Laplace's equation in rectangular coordinates (in sub-section 33.1.2), we consider a metal plate sandwiched between two layers of insulating material so that heat does not flow perpendicular to the plate. In addition we assume a steady-rate temperature distribution.

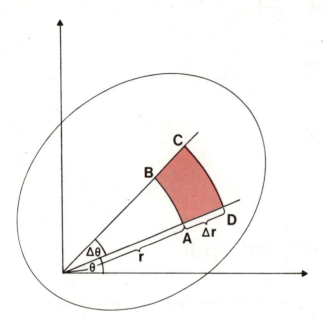

We apply the heat conservation law to a portion $ABCD$ of the plate like the one shown in the diagram. This particular shape $ABCD$ is chosen because it is convenient to the coordinate system, just as the rectangle is suitable for Cartesian co-ordinates. Applying Fourier's Law we obtain for the flow across AB (in the direction of r increasing)

$$J_{AB} \simeq -k_1 AB \frac{\partial u}{\partial r}(r, \theta + \tfrac{1}{2}\Delta\theta).$$

But the length of arc AB is $r\Delta\theta$, and so

$$J_{AB} \simeq -k_1 r \Delta\theta \frac{\partial u}{\partial r}(r, \theta + \tfrac{1}{2}\Delta\theta).$$

Similarly we obtain for the flow across CD (in the direction of increasing r)

$$J_{CD} \simeq -k_1(r + \Delta r)\Delta\theta \frac{\partial u}{\partial r}(r + \Delta r, \theta + \tfrac{1}{2}\Delta\theta).$$

To apply Fourier's law to the sides AD and BC we note that in all cases up to now the heat flow across a line has been

$k_1 \times$ (length of line) $\times$ (the rate of change of temperature with distance measured at right angles to the line).

A small change ϕ in θ corresponds to a change $r\phi$ in AB. So the rate of change of temperature with distance perpendicular to AD is

$$\lim_{\phi \to 0} \frac{u(r, \theta + \phi) - u(r, \theta)}{r\phi} = \frac{1}{r}\frac{\partial u}{\partial \theta} \quad (r \neq 0).$$

For the sides AD and BC Fourier's law therefore takes the form

$$J_{AD} \simeq -k_1 \, \Delta r \, \frac{1}{r} \frac{\partial u}{\partial \theta} (r + \tfrac{1}{2} \, \Delta r, \theta)$$

$$J_{BC} \simeq -k_1 \, \Delta r \, \frac{1}{r} \frac{\partial u}{\partial \theta} (r + \tfrac{1}{2} \, \Delta r, \theta + \Delta \theta).$$

The second of our empirical laws tells us, as before, that for a steady state there is no accumulation of heat, and this, combined with the last (the conservation of heat) tells us that

$$0 = \text{input} - \text{output}$$

$$= (J_{AB} + J_{AD}) - (J_{BC} + J_{CD})$$

$$= -J_{CD} + J_{AB} - J_{BC} + J_{AD}$$

$$\simeq k_1 \, \Delta \theta \left[(r + \Delta r) \frac{\partial u}{\partial r} (r + \Delta r, \theta + \tfrac{1}{2} \, \Delta \theta) - r \frac{\partial u}{\partial r} (r, \theta + \tfrac{1}{2} \, \Delta \theta) \right]$$

$$+ k_1 \, \Delta r \, \frac{1}{r} \left[\frac{\partial u}{\partial \theta} (r + \tfrac{1}{2} \, \Delta r, \theta + \Delta \theta) - \frac{\partial u}{\partial \theta} (r + \tfrac{1}{2} \, \Delta r, \theta) \right].$$

Dividing by $\Delta r \, \Delta \theta$ and taking the limits of small Δr and $\Delta \theta$, we obtain

$$0 = k_1 \frac{\partial}{\partial r} \left(r \frac{\partial u}{\partial r} \right) + k_1 \frac{1}{r} \frac{\partial^2 u}{\partial \theta^2}$$

which reduces to

$$\frac{\partial^2 u}{\partial r^2} + \frac{1}{r} \frac{\partial u}{\partial r} + \frac{1}{r^2} \frac{\partial^2 u}{\partial \theta^2} \qquad (r \neq 0)$$

which is Laplace's equation in polar coordinates.

LINEAR MATHEMATICS